Grapes, indoors and out

Grapes, indoors and out

A Wisley handbook

Harry Baker,
Fruit Officer

Ray Waite,
Glasshouse Superintendent
Royal Horticultural Society Garden, Wisley

Cassell

The Royal Horticultural Society

Cassell Educational Limited
Artillery House
Artillery Row
London SW1P 1RT
for the Royal Horticultural Society

First published 1979
New edition, fully revised and reset 1985
          second impression July 1985
              third impression September 1987

British Library Cataloguing in Publication Data

Baker, Harry
   Grapes indoors and out. — New ed.
   1. Viticulture
   I. Title   II. Waite, Ray
   634′.88   SB397.G7

ISBN 0-304-31088-3

Pest and disease sections
contributed by Audrey V. Brooks,
plant pathologist, and
Andrew Halstead, entomologist,
RHS Garden, Wisley

Line drawings by Peter Mennim
Photographs by Harry Baker, Wilf Halliday, Martyn Rix,
Harry Smith Collection, Michael Warren,
CRH Photographic
Design by Kwan Brenchley

Typeset by Georgia Origination Ltd., Formby
Printed in Hong Kong by Wing King Tong Co. Ltd

Contents

Vines under glass

During the Victorian heyday of the private estate and even to some extent into the early part of the twentieth century, vines were grown on their own in special greenhouses, called vineries.

In some large privately owned gardens there were even separate greenhouses for early, mid-season and late fruiting cultivars, but it was more usual to divide the greenhouse into sections each with a different growing temperature.

After the war the days of the specialised vinery were ended and relatively few people grew vines under glass. Recently, however, interest has re-awakened and now more amateur gardeners are finding that they can grow a worthwhile crop of grapes in the roof space above other plants, even in a small greenhouse. Growing vines in pots can be even more effective, a small greenhouse 8 feet × 6 feet (2.4 × 1.8 m) giving good results.

Most dessert grapes need the protection of glass to ripen properly and achieve their full potential. Certain cultivars, notably the Muscats, need artificial heat in the spring particularly at flowering to aid pollination, and also in the autumn, so that the grapes attain their best flavour. In view of the cost of a green-house, the cost of fuel and the amount of detailed work necessary to keep the vines under control, it is better to use the greenhouse for dessert grapes rather than wine grapes which can be grown outside. The limiting factor in a small greenhouse is head room, but the current fashion is for a greater height to the eaves, so that problem may now be less significant.

The ideal structure for grapes is a lean-to greenhouse facing south; the roof rises across the whole width of the house affording as large an area as possible for training the vines, and in addition the rear wall collects a reserve of heat on a sunny day which is released into the structure at night. However, vines can also be successful in free-standing span greenhouses.

Vines can be grown with or without heat. The latter provides a later crop, but limits the number of cultivars that can be success-fully ripened. In a heated greenhouse the vine can be started into growth in mid-February, and ripe fruit expected in July and August, depending on the cultivar grown. In a cold greenhouse, growth will start naturally in March, with harvesting in late August and into September.

Opposite, 'Mrs Pearson' (see p. 28).

A vine grown without heat can be planted out of doors and the main stem (called the rod) trained through a hole made low down in the greenhouse wall, or it can be planted inside with the roots allowed to grow out in the soil outside.

Vines grown in heated greenhouses are best planted in a specially constructed border. With roots inside, the grower has greater control and success or failure depends on his or her knowledge and judgement. The method is of particular advantage in the colder parts of the country where an early start is required or if there is any risk of the roots growing into areas which might become waterlogged in winter. In the past it was always said that for grapes started very early in the year the inside border ensured that the root action was on the move at the same time as the warmed top growth. However, this is a counsel of perfection and for most needs to-day outside planting will be satisfactory making for easier initial preparation and less frequent attention to watering and feeding.

PREPARATION FOR PLANTING

Whether planting is to be done outside or inside the greenhouse, soil preparation means double digging, incorporating well rotted manure or compost, together with a dressing of general fertilizer such as Growmore. Bonfire ash can also be added if available. If there is any suspicion that the soil might become water-logged construction of a special bed will be necessary. Make the border $2\frac{1}{2}$ to 3 feet deep (75–90 cm) with solidly constructed sides and base, for example brick retaining walls and a concrete floor. The floor should slope from back to front and from one end to the other and a land drain pipe 4 inches in diameter (10 cm) laid along the lower edge with an outlet at the lowest corner leading to a ditch, surface water drain, or a soak-away prepared by digging a large hole and filling it with stones or brick rubble. Spread similar drainage material to a depth of 6 inches (15 cm) over the whole of the base before filling up the border space with soil. This soil should be capable of holding reserves of water and plant food, at the same time being open and porous. The recipe for John Innes Potting Compost No. 3 (7 parts by volume of sterilised loam or fertile garden soil, 3 parts by volume of peat, 2 parts by volume of coarse sand, plus 12 oz (300 g) of John Innes Base fertilizer and $2\frac{1}{4}$ oz (55 g) of ground chalk to every bushel (36 litres)) will provide a satisfactory medium. If wood ash from a bonfire containing charcoal is available, this can also be included in the mixture.

The young plant will not require a very extensive root-run in the first year, and the old practice of making up a narrow border for

As the bed is enlarged in the first years of growth, good drainage
material is placed below the compost mixture.

the first year and adding to the width annually has much to
commend it, for it makes less initial work and reduces costs. The
method ensures that a medium full of nutrients and in good
physical condition is being added year by year during the early
life of the vine. Concrete building blocks provide a convenient
means of retaining the compost, but ordinary house bricks or
planking will serve equally well.

The growing vine will need to be trained inside the greenhouse
so preparation must include the fixing of wires for support. It is
usual to run the wires along the length of the structure starting at
about 3 feet (90 cm) off the ground, and continuing at about 9
inches apart (22 cm) and 12 to 18 inches from the glass (30–45 cm).

In a small greenhouse it is only feasible to allow 12 to 14 inches
(30–35 cm) from the glass, but the young shoots tend to grow
vertically and quite rapidly and will soon stub themselves against
the glazing and become scorched or distorted if they do not have
enough space. It is also essential to keep the air moving between
the glass and foliage so the greater the space given to the wires the
better. In a larger greenhouse, allow about 18 inches (45 cm).

PLANTING

In the small greenhouse there will probably be room to plant only one vine. This is best done at the opposite end to the door so that the rod can be trained along the whole length of the ridge. Alternatively, the young vine can be planted to one side, at the end near the corner, and run along the side wall.

Traditionally vines were planted side by side and this method lends itself perfectly for the large greenhouse especially of the lean-to design. Plant the rods at 4 feet apart (1.2 m) as the spread of lateral growth can be expected to reach 2 feet or more (60 cm) either side.

Young vines are usually sold in containers and therefore can be planted at any time of the year, but November and December, when the plants are dormant, are the best, as at that time the main stem can be cut back by two-thirds without fear of excessive sap 'bleeding'. This severe pruning ensures vigorous growth in the summer and also stimulates the lower buds to form side shoots (laterals), so encouraging lower spur systems to be formed. If plants are not obtained until later in the winter, it is unwise to prune them, otherwise the rising sap will bleed for several days from the cut, and considerably weaken the vine. Similarly, vines planted in full growth should not be pruned until after leaf-fall.

It is not always possible to purchase strong vines (stem about $\frac{1}{2}$ inch in diameter: 1 cm) in which case it is better to pot on into a larger pot, reduce the main stem to one-third its original length, and allow the plant to grow for another season before planting in the following winter.

THE FIRST YEARS

In the first year after planting (and after the initial hard pruning) allow the main stem to grow unchecked so that it produces the maximum length of vigorous shoot. If the roots are growing well, the main shoot should reach 10 feet (3 m) or more and produce lateral shoots over much of its length. During the summer these shoots are cut off when they have made five or six leaves, and any sub-laterals (side growths formed on the laterals) produced are cut to one leaf. Both main stem and laterals are loosely tied into position on the wires as they grow, allowing the stems ample room to increase their girth.

At the end of the season and as soon as the leaves have fallen, the main stem is cut hard back again, removing two-thirds of the summer's growth. The lateral shoots are pruned to one plump bud.

In following years the same technique is carried out until the main stem, after being pruned by two-thirds of its length, has reached the furthermost wire. Ultimately this form of training will result in what is known as the rod and spur system, a permanent framework consisting of a main stem with the laterals forming spurs at intervals of 9 to 12 inches along it (22–30 cm). It is from these spurs that the fruiting shoots arise each year. The main stem between the spurs is barren. This is the simplest form of training, but it is also possible to grow two or more rods on one plant or allow the rod to branch and extend until one vine fills an entire greenhouse, as does the famous vine at Hampton Court.

The glasshouse vine, grown on the rod and spur system of pruning before (left) and after (right) pruning immediately after leaf fall.

WINTER PRUNING OF ESTABLISHED VINES

Vines should be pruned immediately after leaf-fall, which is usually in late November or December. At this time the greenhouse must be quite cold in order to keep the vines dormant. This is done by giving as much ventilation as possible after harvest so that the current year's growth is fully ripened. This hardening of the growth will, with the early winter pruning, avoid the real risk of a harmful level of bleeding when the sap rises in the spring. Bleeding can be prolonged and abundant, and although various 'cures' have been suggested, none are wholly satisfactory. If later pruning is unavoidable, the application of carpenter's knotting to the cuts immediately after pruning may help, but it is better to avoid the trouble rather than to cure it.

With the established rod and spur system, winter pruning consists of cutting back all growth made in the past year (other than that required for extension or replacement) to within one or two buds of the older wood. One bud is enough if it is strong and plump, but often the basal bud is small and unlikely to give fruiting wood. Pruning to two buds will give two growths so that should one become damaged or broken away, there is still one to be tied in to fill the space which would otherwise be left.

Indoor vines can also be grown on a replacement pruning system. This requires more skill but will generally result in higher quality fruit. With this system the winter pruning is much the same as that for the Guyot method of growing vines in the open (see p. 38). Two new main stems are allowed to grow each year, one of which is retained for fruiting and the other cut back to two buds to produce the two rods for the following year. The rod which has fruited is cut out completely.

WINTER MAINTENANCE

The practice of scraping away the oldest bark from a rod takes time and some growers regard it as unnecessary; but if there is time, it is still well worth doing. Special attention needs to be given to the flaky bark on the spur, but take care near the buds so as not to damage them. Much of the fibrous bark can be rubbed or pulled away but when scraping with a knife, the green tissue must not be exposed, going down just to the smooth brown bark.

By scraping away the old bark, over-wintering pests, such as red spider mite and mealy bug, are either removed or exposed to a pesticide which can be sprayed on, but preferably brushed on ensuring that it is rubbed into crevices and angles of the bark.

If vines are growing in the greenhouse on their own, a 4% solution of tar oil winter wash (30 cc of tar oil in $\frac{1}{4}$ litre water: 1 fluid ounce to $1\frac{1}{4}$ pints) may be applied. The smell, although not harmful, can be persistent; if this is offensive near the house, an alternative is malathion at the strength recommended for spraying. It is best to complete this work by early January, so that there is no chance of injuring the buds. This cleaning of the rods is more convenient if they are untied and lowered from the wires. Remove all the old ties from the wires as these too may harbour pests.

The next job is to renovate the inside bed. The remains of the old mulch are raked from the bed and the surface soil gently pricked over with a fork. This loose soil is raked or brushed off with a stiff broom or besom. The roots thus exposed are immediately top-dressed with a good loam-based compost such as John Innes No. 3.

No more needs to be done for the next two or three months, until the vine is started into growth. As this time approaches, the soil in the bed is watered thoroughly; a second watering a week later may be needed if the root area has become very dry during the winter. After watering, a mulch of well rotted farmyard manure, spent mushroom compost or similar material should be applied; this will help to regulate the soil moisture during the growing season. This is also the time to extend the vine border in width as necessary.

For grapes planted outside the greenhouse, no special winter treatment is required other than an application of a good general fertilizer, when growth begins.

THE SPRING ROUTINE

If warmth is required for the needs of other plants, heat may be provided in February; the middle of the month is an ideal starting time for vines.

The rods which have been loosely tied to the wires for the winter operations, are untied and lowered so that when supported about a third of the way up, the end is bent down, almost touching the ground. In this way the rising sap is prevented from rushing to the top buds and stimulating them into growth at the expense of those lower down. When growth has started on all the spurs, the rods are tied into their summer positions on the wires.

Temperatures at this stage may rise rapidly in sunlight and as a gentle start is required, careful ventilation must be given at about

'Black Hamburgh' has a good flavour and does well in pots.

19°C (66°F). With artificial heat a night temperature of 4 to 7°C (39–45°F) should be the aim. Keep the atmosphere moist by damping down the floor and walls on fine days. Until flowering begins, the rods can be sprayed overhead with clean water, but do this in the morning before the sun becomes too hot, for there can be a risk of scorching the young foliage by the action of the sun's rays through the water droplets.

If it is necessary to reduce the number of growths to give one shoot on a spur, the surplus ones should be removed as soon as possible. In practice, it is safer to pinch back the weaker unwanted shoots to two or three leaves, in case a potential fruiting shoot is accidentally broken when tying in.

As the shoots will grow vertically, they must be brought down to the wires; this can be rather tricky as they may easily snap at the junction with the older wood. Growths should be trained down gradually and this will need to be done every other day. It is unwise to tie in while the growth is very young and brittle, and better to wait until a little fibrous tissue has formed. The task should be done in the morning, for although later in the day the shoots can be limp and more easily managed, at night they become turgid again and if tied too tightly may break. As the light intensity increases, apply light shading to the glass to prevent leaf scorch. This may need to be done as early as April with a small greenhouse. Shading will also help to reduce excessively high and too rapid rises in temperatures and thus to some extent prevent drying conditions.

Removing a surplus shoot from a spur (left), and a lateral shoot tied down with a running noose (right).

A stopped shoot showing the formation of laterals.

FLOWERING AND POLLINATION

By the time the shoots are 18 to 24 inches long (45–60 cm), the flower trusses will be quite prominent and a start can be made in pinching out the shoot tips. Two leaves are allowed to form beyond the flower bunch before the growing tip is removed. If no bunch has been formed, the shoot is left to make 6 or 8 leaves before pinching. All side shoots that form on the new growth, should be pinched at one leaf.

Although some cultivars set their fruit very readily, it is unwise to leave pollination to chance. With freely setting cultivars like 'Black Hamburgh', 'Foster's Seedling' and 'Alicante', all that is needed is to give a brisk shake to the rods. For shy setters like the Muscat types, the easiest effective method is to stroke the bunches gently with cupped hands in order to transfer the pollen from the stamens to the receptive stigmas. If pollen can be transferred from a different cultivar, so much the better. Always try to pollinate at mid-day. At this critical stage, keep the atmosphere on the dry side, and although warmth will help pollination, air must be kept circulating. Even on a cool day, an inch or two of ventilation will help. Do not spray overhead on such days.

A shoot showing flower buds and lateral growth.

THINNING THE CROP

If all the bunches are allowed to mature, the quality of fruit will suffer: in addition the vigour of the shoot growth is lost and fruiting may be upset for several years. The number of bunches left to fruit must therefore be regulated; and guidance on this is given in the notes on different cultivars on pp. 26–28. As a general rule of thumb, a pound of fruit is retained for every foot run of rod (450 g per 30 cm). This lightening of the load should be done as soon as it is possible to assess which bunches show promise of being shapely, well set, of good size and in such a position that they will develop freely.

Next the grapes must be thinned. It is not a difficult task, but takes time. Special grape thinning scissors are obtainable but ordinary small scissors will do, provided they are sharp, particularly at the tip, and are not too cumbersome. A small stick about 6 inches long (15 cm), forked at one end, is helpful in steadying the bunch. Grapes should never be touched with the fingers, as this damages the bloom on the skin. This damage not only disfigures the bunch, but removes the extra protection given by the water repellent qualities of the bloom.

17

Above: bunches before (left) and after (right) first thinning.
Below: bunches before (left) and after (right) second thinning.

Occasionally it is necessary to shape the bunch before thinning, and this usually involves the removal of a straggling arm at the top of the bunch. Any large bunches may need to have their top branches, referred to as shoulders, supported by looping them up to the wires with raffia. Excess grapes in the middle of the bunch are cut away first, then any undersized berries as these are probably imperfectly set. Next, the outer berries are spaced to allow room for swelling. The shoulders should not be thinned too much as they should stand out firm and square when the bunch is harvested.

Thinning can seldom be completed in one operation and it is usually necessary to remove a few more grapes three or four weeks after the first thinning. A final thinning may be required after the pips have formed when the grapes undergo a second period of swelling.

WATERING AND FEEDING

A vine in full leaf will make heavy demands on soil moisture in warm weather. The intervals at which water is required will vary according to the climate and soil conditions. If the vine's roots are confined inside the greenhouse, a generous application of water will probably be necessary every 7 to 10 days, and provided drainage is good, the risk of over-watering will be slight. It is better to water thoroughly so that moisture penetrates several feet into the border soil. If the roots are outside, they may be supplied by natural rainfall for much of the year, but should not be forgotten in prolonged dry spells.

Feeding will be required while growth is active. For good quality fruit the main need is for a high potash feed, and any of the tomato fertilizers available will be ideal. Water on a solution at two- to three-week intervals from about a month after starting into growth until the fruit begins to ripen.

Sometimes the vine is not vigorous – a condition brought about perhaps by overcropping, old age or poor soil conditions, in which case the application of a general feed, e.g., with equal quantities of nitrogen, phosphate and potassium, or one containing higher nitrogen will be beneficial, but these should only be used from April to May.

RIPENING

Throughout the summer and autumn a healthy vine will grow vigorously. Any new shoots not needed to fill gaps in the foliage canopy or to add to the length of rod for the following year are cut back after one new leaf has formed; this must be done promptly or the growths will rapidly get out of control. Remove tendrils immediately they form.

When the grapes start to change colour, there is a greater risk of splitting. This stage is easily recognised in black or tawny grapes, but is not so apparent with green ones, although a slight change of colour towards white or amber, is noticeable.

A damp atmosphere encourages splitting. Damping down of the path and other surfaces should be done by the middle of the day so that any excess moisture has had a chance to dissipate before night. If there are other plants in the house, they should also be watered well before evening. Leaving the lee side top ventilator open for an inch or two will help to dry the atmosphere and will also control to some extent any early build up of heat in the morning. The air temperature should never be allowed to rise rapidly, because the temperature of the grape rises slowly, and moisture from the air will condense on the fruits which may then split. If the soil is allowed to get very dry, the fruit skins will harden and when water is finally given, they are likely to split. A thin mulch of straw applied over the soil when the fruit has formed will help to keep the soil uniformly moist and the air in the greenhouse on the dry side. From the early stages of ripening, look over the bunches two or three times a week for diseased grapes, removing those with a split skin: continue this until the end of the season. Cut out the bad grapes using scissors and the forked stick, working with care so as to keep the bloom intact.

Grapes are not completely ripe when fully coloured and require 'finishing'. This means allowing them to hang for a period so that sugars are formed. The finishing period varies with the cultivar and time of year. 'Black Hamburgh' and 'Foster's Seedling', ripening in the summer, need only two or three weeks to mature fully; 'Muscat of Alexandria', 'Mrs Pince' and 'Syrian', which ripen later, need to hang for eight to ten weeks before they are at their best.

When harvesting the grapes, handling will be made easy if the branch carrying the fruit is cut with a 'handle', i.e. leaving about 2 inches (5 cm) on either side of its join with the main stem. This handle allows the bunch to be inspected, carried, mounted on a showboard or placed in a dish without disfiguring the fruit.

Ripe grapes may be kept in good condition on the vine for several weeks by using artificial heat and ventilation to maintain a cool, dry atmosphere (about 7°C:45°F).

If a cool, even temperature can be provided in a room, it is worth cutting the late grapes with a stem in December. The stem can be placed in a jar of water and in this way fruit will keep for a considerable time. 'Lady Downe's Seedling' is particularly good for this purpose.

Opposite, above: 'Black Hamburgh' grapes in a greenhouse.
Opposite, below: 'Lady Hutt'.

Vines in pots

The cultivation of vines in pots has several advantages. The containers can be readily moved within the greenhouse and moved outside after fruiting. Such outdoor treatment will ripen the shoots and give the cold environment necessary for satisfactory dormancy. Clay pots will need to be plunged to reduce the chance of frost damage to the containers, but whatever the receptacle, plunging gives stability in windy weather.

Much of the work described for vines under glass, applies equally to the management of pot-grown grapes except that they are grown as standards, with a long stem producing a 'head' of shoots. The method deserves to be more widely used because worthwhile crops can be obtained, especially by owners of the smaller greenhouse. Where space is limited, more than one cultivar can be grown and with care, the more usual crops can be cultivated at the same time. The vines are taken out of the greenhouse in the autumn, leaving room to house chrysanthemums, for example. Heating will give a long season of growth and therefore earlier ripening, but an unheated greenhouse can also give good results.

Early fruiting cultivars will be found to be the most satisfactory with 'Black Hamburgh' still possibly the most suitable; but 'Buckland Sweetwater', 'Foster's Seedling' and the Frontignan types all give good results.

Well established rods should be obtained and are best potted just as the roots become active in the spring. Weak rods will need to be grown on so as to build up the main stem and it will be necessary to cut the rod back to one bud so as to induce vigorous growth during the summer. Any pruning should be done during November and December. Rods with a diameter of $\frac{1}{2}$ inch or more (1 cm) can be pruned to a bud (eye) at 3 to 5 feet above the pot (0.9–1.4 m). The length of stem chosen will depend on the height of the greenhouse.

Vines may need to be potted into a larger size pot as the root system grows, and this is best done as the roots start into active growth in the spring. A vine in a 5-inch pot (13 cm) should be potted on into a 7-inch pot (18 cm) but plants with larger root balls can be moved into a 9-inch pot (22 cm). A loam-based compost, e.g. John Innes No. 3, is best, because it will ensure that the plant will not become top heavy.

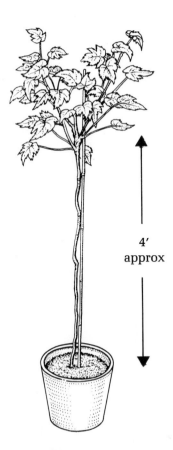

4′
approx

A vine in a pot, showing the length of stem before the branches arise.

Once started into growth, the shoots soon lengthen and the top four or five shoots should be left to develop a 'head', being pinched when they have made five or six leaves. Any growths on the stem should be removed completely. Flower clusters are unlikely to appear but if they do, remove them, as they should not be allowed to develop at this early stage.

For the second year the ripened shoots should be pruned in the dormant season to one bud. This is the start of building up the spur systems. These will form the permanent head of a pot-grown vine. As active growth begins in the spring, scrape off the existing top layer of soil then add a top dressing with John Innes Potting Compost No. 3 to the top of the pot, being careful to leave enough depth for thorough watering. To encourage strong growth and to ensure that it is well ripened stand the vine out of doors from mid-summer onwards.

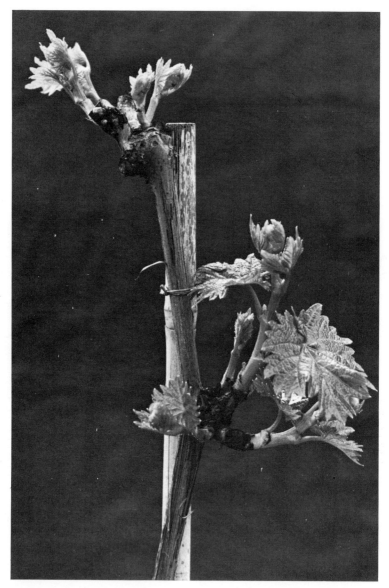

'Buckland Sweetwater' showing the new shoots arising from multiple buds.

During the third year prune hard to one good bud in early winter. A move into a larger container will almost certainly be required, but this is likely to be the last move as with top dressing and feeding, a vine can remain in the same pot for many years.

The vine can produce its first fruit in the third year. Pinch the growths at two leaves beyond the bunch, and any laterals and sub-laterals at one leaf.

The growths will need support and this can be done by tying them to a central cane in a maypole fashion. A general liquid fertilizer is best applied every 7 days up to fruit set, after which a high potash feed (e.g. a tomato fertilizer) should be used.

In warm situations, it may be possible to finish ripening the fruit outside, but it will need to be protected from damage from birds and wasps, by covering with netting, or enclosing the bunches in nylon stockings.

Branches are secured to the central cane with raffia or garden twine.

Cultivars for growing under glass

(including pots)

Greenhouse grapes grown to-day can be grouped as follows:

SWEETWATER (S) These are early grapes, quick to mature and ripen. They are ideally suited for an unheated greenhouse. They are very sweet and juicy but without the muscat flavour. They have very thin skins when at their best but do not hang on the vine in good condition for long.

MUSCATS (M) are the finest grapes for flavour, but they need heat. They are the second group to mature and they hang well in good condition if a little warmth is provided. The fruit is firm and luscious and if allowed to hang until the skin starts to wrinkle the berries often take on the flavour of raisins. They are unfortunately difficult to get to set well unless they have warmth and some hand pollination.

VINOUS (V) These are the really late grapes and in the past were grown to provide fruit well into the new year. They are usually strong growing and crop well but need to hang for a long time in a warm greenhouse to develop their best flavour. They are not a worthwhile proposition unless heat can be provided for them in the early winter months.

Alicante (V) A black (late) grape. Vigorous grower. Fruit sets freely and needs early and severe thinning. It is inclined to form an unshapely bunch but when well thinned it can produce a handsome and imposing bunch for show with its large, almost round, jet black berries covered in heavy blue/grey bloom (9 medium size bunches/12 feet (3.6 m) of rod).*

Black Hamburgh (S) has the best flavour of this group and is justifiably the grape best known and most widely grown under glass in this country. Excellent for pots. Of good constitution, it sets fruit freely and can be ripened well in an unheated greenhouse. If allowed to hang too long after ripening the skin becomes very thin and easily broken and the fruit deteriorates. (12 medium-size bunches to 12 feet (3.6 m) of rod).

*These figures are given as a guide to a reasonable number of bunches for a healthy vine. (N.B. a medium sized bunch weighs about 1 lb (450 g).).

Buckland Sweetwater (S) A round white grape carried in short broad bunches. It sets freely and ripens early particularly as a pot-grown vine. It has a pleasant flavour when in good condition but the skin thickens and the fruit deteriorates if it is allowed to hang too long on the vine. It is not a vigorous cultivar and needs to be fed well for good results. (12 bunches to 12 feet (3.6 m) of rod).

Canon Hall Muscat (M) is an almost round white grape which becomes pale amber on ripening. Of excellent flavour when well grown, but it is difficult unless heat is provided at flowering time and for ripening the fruit. Berries very large. (10 bunches/12 feet: 3.6 m).

Frontignan (M) Both the black and the white Frontignan grapes are early maturing with small grapes of good quality. Well suited to growing in pots under glass. (14 bunches/12 feet: 3.6 m).

Foster's Seedling (S) The best of the white sweetwater grapes. Useful in pots. It sets freely, the bunches are of medium size and shapely, the grapes oval, juicy and of a pleasant flavour, ripening early. They should be eaten fairly soon after ripening or they will lose flavour. It has a good constitution and grows well in an unheated greenhouse. (12 bunches/12 feet: 3.6 m).

Gros Colmar (V) A strong growing, round, black grape, with very large fruits and handsome bunches. The skin is thick and the flavour poor until the fruit has had a long period of ripening in a warm greenhouse through the early winter. (10 bunches/12 feet: 3.6 m).

Lady Downe's Seedling (V) One of the better flavoured vinous grapes with round black berries carried in long tapering bunches. It is of good constitution but requires warmth at flowering time to ensure a good set of berries and in the early winter to ripen its fruit. (10 bunches/12 feet: 3.6 m).

Lady Hutt (S) is a round white grape ripening several weeks later than others in that group. It is a robust grower, sets freely and produces large bunches of thin-skinned juicy berries. (5 bunches/12 feet: 3.6 m).

Madresfield Court (M) is an early ripening oval, black grape of high quality. Its berries are large and covered with dense blue/grey bloom, the flesh firm but juicy and of good flavour, but the skin is tough. The berries are liable to split on ripening and for this reason some growers used to leave top shoots to run on without stopping at ripening to reduce sap pressure. (10 bunches/12 feet: 3.6 m).

27

Mrs Pearson (M) is a round white grape which ripens late. It grows strongly, sets freely, bears fruit of excellent flavour but with thick skins and will hang in good condition after ripening. It does however need warmth throughout the ripening and holding period. (10 bunches/12 feet: 3.6 m).

Mrs Pince (M) is a strong growing, oval, black grape which produces late grapes of excellent quality but requires warmth and hand pollination at flowering to obtain a good set of fruits. It also requires warmth throughout the early winter when it is ripening. (10 bunches/12 feet: 3.6 m).

Muscat of Alexandria (M) When well grown there is no better flavoured grape. It has a good constitution but needs warmth and hand pollination if it is to set a good crop of oval white grapes. Warmth is also needed to ripen the fruit although if heat is used in the greenhouse in spring to start the vine into growth early the grapes should ripen in the September sunshine. (10 bunches/12 feet: 3.6 m).

Muscat Hamburgh (M) An oval, black grape of excellent quality which ripens before most of the other muscats. It is sometimes recommended for growing in an unheated house but it is difficult to grow well. It needs warmth and hand pollination at flowering time (cross pollination, using pollen from a different cultivar is helpful). Although it is fairly vigorous in growth it has a strong tendency to shanking (see p. 64). (9 bunches/12 feet: 3.6 m).

Syrian (V) A late, oval, white grape of great vigour. It will rapidly get out of control unless regularly pruned in summer. It carries impressively large bunches but the flavour is poor unless they are allowed to hang in a warm house until Christmas. (5 bunches/12 feet: 3.6 m).

Trebbiano (V) An oval, white late grape similar in many respects to 'Syrian'. It is notable for the size of the bunches, one of which is on record as weighing 26 lb 4 oz (12 kg). (5 bunches/12 feet: 3.6 m).

Propagation

Vine cuttings are easy to root. Soft, green cuttings taken in summer need some heat for rooting but hardwood cuttings taken in winter can if necessary be rooted outside.

Material for hardwood cuttings can be made from the prunings in December. If they are to be rooted in the open, cut the prunings to about 8 inches long (20 cm), with three buds. Make a cut above the top bud and another just below the lower bud before inserting the cutting to a depth of about 6 inches (15 cm) in sandy well-drained soil. Those cuttings rooted in the open garden should remain undisturbed until the following autumn.

Rooting is also possible in containers in a cold frame; a cutting of two buds will be much more convenient to insert in a pot. If cuttings of this type are inserted singly in 3½-inch pots in a compost consisting of two parts sand, one part loam and one part peat in January they should be well rooted by early June. They may then be potted on into 5-inch pots in John Innes No. 2 potting compost.

With more sophisticated apparatus such as a heated propagating case, a single bud cutting is inserted in January in a 2½-inch pot in a compost of equal parts of peat and sand and placed in the case with bottom heat of 18°C (64°F). These cuttings are made with a cut just above the bud and another about 2 inches below (5 cm). They are pressed vertically into the surface of the rooting mixture with the bud just clear of the surface. Other methods, using a horizontally inserted single bud cutting may be found in old literature but are no more effective.

Green softwood cuttings may be made in July or August, using shoots of the current year. Lengths of stem about 4 inches long (10 cm) are cut just below a bud, to include either the growing tip or at least two buds. These are inserted in a mixture of equal parts of peat and sand in a 2½-inch pot; in a propagating case with a bottom heat of 18°C (64°F) they will root readily in six or eight weeks. The first move should be into 3½-inch pots of John Innes No. 1 potting compost and then when the plants are sufficiently robust they should be potted on again into 5-inch pots and John Innes No. 2 potting compost. In their first year the young vines, although perfectly hardy, will make better growth if kept in a cool greenhouse.

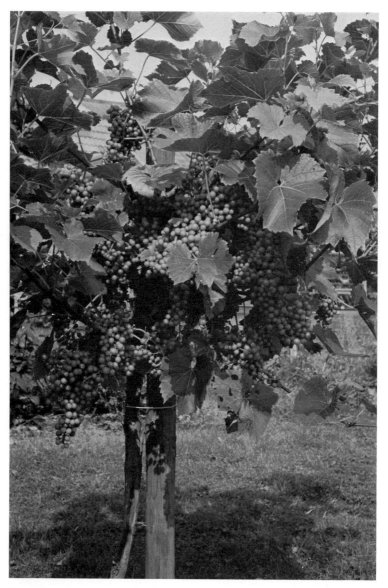

'Seyval', grown as a standard.

Vines can be grafted and some commercial growers of wine grapes in this country take the precaution of grafting their vines onto rootstocks which are resistant to attack by a root pest, *Phylloxera vastatrix*, which has over the years been a major worry to owners of continental vineyards (see p. 57).

Vines in the open

The vine is a sun loving plant and happiest outside on a warm wall; nevertheless, grapes can be grown out in the open, at least in the south, and further north where there is a favourable micro-climate, provided the right cultivars are chosen. Thanks originally to the work of a few pioneer viticulturists, principally Mr Barrington Brock in the 1950s, a number of grapes have proved reasonably successful outside and more have been introduced since. However, even with the correct choice of cultivar, grapes will only fruit successfully and the berries acquire their necessary sugar content and flavour for good wine or dessert in warm, dry summers, and of course without frost at flowering time. They will not do well in areas of high summer rainfall, strong winds or where the climate is cool during the growing season.

In general, outdoor grapes can be planted south of a line from the Wash in the east to south Wales in the west. Further north than this, it is best to grow the vine on a wall or provide some kind of protection and even in the south the use of glass or plastic to protect against frost and as a means to increase the temperature is very beneficial.

In essence, the United Kingdom is a marginal area for viticulture and whilst there will be vintage years for the wine maker, there will be far more when the resultant wine might be regarded even by the best of friends as *vin ordinaire*!

CHOOSING THE SITE

With vines grown in the open the choice of site is most important. It must be sheltered and yet in full sun and at an elevation of not more than 300 feet (90 m). The ideal site is a south or south westerly slope but level ground, or even a site sloping slightly towards the east or north is satisfactory. Avoid planting in a frost pocket as frost in the late spring can damage the blossom and young growth, possibly wiping out the potential crop for that summer. In situations where the gardener has no alternative, he should be prepared to protect the vines in some way, for example by covering them with hessian, when frost is expected.

Grapes on walls and fences are in the best position because of the reflected warmth from the structure. The position must be

31

sunny and sheltered, however, and ideally one with a southerly aspect, though a west facing support is acceptable.

SOIL PREPARATION AND PLANTING

Vines are tolerant of a wide range of soils provided they are at least 12 inches deep (30 cm) and well drained. Light to medium loams are ideal but very fertile land might cause problems with excessive vigour, and extremely shallow soils over chalk with lime-induced chlorosis and poor growth in very dry seasons. Acid soils require to be limed to bring the pH up to between 6.5 and 7.0.

Grapes root deeply and will not tolerate badly drained land nor a hard impenetrable pan beneath the surface. Heavy clay, where there are drainage difficulties, should be avoided but if this is not possible then some kind of drainage system must be installed as the first stage of soil preparation. The type depends upon the scale of planting. In a small area, where only one or two vines are planted, a simple soak-away about 2 feet square by $2\frac{1}{2}$ feet deep (60 cm x 75 cm) constructed of broken bricks, clinker and rubble would suffice, but for larger areas, say for one or more rows of vines, a length of tile or plastic drains or even a herringbone system of drains may be necessary.

The planting site should be prepared well beforehand by double digging to break up any hard layers. Ensure the area is free of weeds, and add lime if necessary. Too rich a soil is undesirable and no bulky organic manure is required unless the ground is poor. Even then only a light dressing of well rotted manure or compost at the bottom of each trench should be applied. In the final preparations, and just before planting, rake in John Innes base fertilizer* or Growmore at the rate of 3 oz per square yard (100 g/m²).

Plant at any time from October to March. Weak vines, of less than pencil thickness, are best over wintered in the protection of a cold frame and planted out when the danger of frost is over. Most plants supplied by nurserymen will be on their own roots, although a number of plants are being imported from the continent which are grafted. The latter are resistant to the vine louse, *Phylloxera vastatrix*, though fortunately this is not a problem in this country. Some rootstocks are more suited to certain soil types than others (see p. 53).

---

*John Innes base consists of 2 parts by weight of hoof-and-horn meal, 2 parts superphosphate of lime, and 1 part sulphate of potash. Growmore has an analysis of 7% N: 7% P_2O_5 : 7% k_2O.

Firm planting is essential and the roots of the young vines (often pot grown) must be carefully spread out to encourage quick establishment. Subsequently mulch them with a little rotted manure or compost so that the lower buds are covered for the winter. With grafted plants the union should certainly be covered in winter. In the spring the mulch should be gently pulled away from the stem. The graft union must be above soil level to prevent scion rooting because if this occurs the desired effect of the root-stock on the vine will be lost and the plant will no longer be resistant to *Phylloxera*.

A pot grown vine, with a good root system, before planting.

The vine in the planting hole, with the roots well spread out.

GROWING ON WALLS, FENCES AND PERGOLAS

The vine is a climbing plant amenable to many forms of training – the cordon, espalier and fan for example. Whatever the method used, it should be grown in such a way that a permanent framework of branches is maintained.

On walls and solid fences, it is usually grown as a vertical cordon (rod) in single or multiple form with each rod spaced $3\frac{1}{2}$ to 4 feet (1.2 m) apart to allow room for the laterals carrying fruit. On a low fence or under a window the vine could be grown as an espalier with each horizontal arm spaced about 18 inches (45 cm) above or below its neighbour. Plant about 9 inches (22 cm) away from the wall or fence to allow room for the expansion of the main trunk.

Support. A wall trained grape requires a support system of wires generally taken horizontally and secured to the wall with lead wall eyes. Galvanised fencing wire, gauge 14, spaced 10 to 12 inches (25–30 cm) apart can be used.

Above: 'Müller Thurgau', a popular variety, which gives an excellent wine.

A row of cordon vines in winter.

Training. The strongest shoot is selected to form the rod and after leaf fall is pruned back to a bud where it has become well hardened and woody. Any side shoots are cut back to leave two buds nearest the base. This leading shoot is extended each summer until the desired length is reached and is shortened as just described each winter. If several rods are required from the same vine the first move must be to establish two horizontal shoots. From these the desired number of vertical rods are then developed in subsequent seasons.

Pruning. Pruning, deshooting, pinching and tying in must be done regularly, and basically the method is similar to that for vines under glass except that the timing of the summer's operations is later.

a) Summer. The first task is to thin out the young laterals growing from the spurs, but not until the flowers are produced. The shoots are then thinned to one per spur or one about every 10 inches (25 cm). Retain those which are carrying the strongest flowers. The unwanted laterals are cut back to one leaf, rather than being removed completely, to reduce the risk of blind spurs. The retained laterals are stopped by pinching out the growing point at two leaves beyond the flower truss. Any sub-laterals subsequently produced are stopped at one leaf. Any barren laterals are thinned in the same way and stopped at six leaves, and their sub-laterals to one leaf.

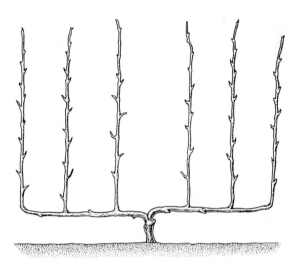

Training a vine against a wall, showing how the fruiting arms are trained.

Left: Standard vine 'Seyval' before spur pruning.
Right: Standard vine 'Seyval' after spur pruning.

(b) Winter. As soon as the leaves are off and before the end of December, the spurs are pruned by cutting back the previous summer's laterals to one or two buds ensuring the cut is made to a plump bud. Such a bud is more likely to produce a shoot with flowers in the next summer. Thin the spurs if necessary to 9 to 10 inches apart (22–25 cm).

Where a new replacement rod has been trained in, cut out the old rod back to the replacement. Paint the cut with carpenter's knotting or a bitumen wound paint. Remove any thin unripe extension growth by cutting back to strong nut brown wood.

The opportunity should also be taken to fill in empty wall space with extension growth or suitably placed laterals. An old rod carrying too many dead spurs can be replaced by leaving in a strong replacement shoot at or near the base. In the autumn the old rod can be cut out, back to the replacement. Tie in new growth where necessary and check any old ties to ensure there is no constriction taking place. The wirework should also be examined, renewed and extended if necessary.

Watering. A vine growing against a fence or wall in the open needs very little watering once it is well established, except in periods of drought. Whilst it is young, however, watering is advisable in the summer in dry periods to promote strong growth which will be better able to withstand the ensuing winter frosts. A vine growing against a house wall where the soil is usually sheltered by the eaves of the house can become very dry at the roots. Grapes grown in hot, dry conditions are more susceptible to mildew. It is essential to keep the soil reasonably moist throughout the summer by irrigation, when necessary. This involves applying about an inch of water (4½ gal. over 1 sq yd) every 7–10 days, in hot dry weather. A mulch of well rotted stable manure or compost applied in late February will help to conserve soil moisture.

Feeding. In late February rake away the old mulch and apply Growmore fertilizer (7%N : 7%P$_2$0$_5$: 7%K$_2$0) or its equivalent at 2 oz per sq yd (66 g/m^2) plus sulphate of potash at ½ oz per sq yd (15 g/m^2) over the rooting area and then apply a new mulch of well rotted stable manure or compost to a depth of about 2 or 3 inches (5–7 cm).

If magnesium deficiency symptoms appear apply magnesium sulphate (see p. 62). With dessert grapes give the vine a liquid feed high in potassium, about ½ gallon per plant every 10 days throughout the growing season, but stop once the berries start to ripen.

Cropping and thinning: bunches and berries

The number of bunches to allow depends upon the age and condition of the vine. As a guide, a well grown mature vine should be able to carry one bunch every 10 to 12 inches (25–30 cm) along the length of the rods. Certain cultivars which produce very small bunches with tightly clustered berries, 'Brant' and 'Cascade' for example, can be allowed more, say, at every 8 inches (20 cm).

Overcropping must be avoided, a young vine in particular will frequently remain barren for a number of years after it has been allowed to carry too many bunches. Only two or three should be left on a 3-year-old vine, perhaps four or five in the following year, and so on, provided growth is sufficiently vigorous to sustain the crop. Overcropping and the removal of too many leaves in the summer will also depress colour, sugar content and flavour.

The thinning of the berries within the bunch is not necessary for wine, but could be done to advantage for dessert, provided they are cultivars which can produce large sized berries, for example 'Siegerrebe' and the Chasselas clones. It is not necessary to thin 'Brant' and similar cultivars, as the berries are naturally small.

GROWING VINES IN A VINEYARD

There are many methods by which a vine can be trained in the open. It could for example be grown in a sunny corner as a vertical cordon trained up a stout pole or as a standard like a rose. It could also be grown in a container (see pp. 22–25). The grape in any of these examples would be spur pruned in the winter and the laterals pinched back in the summer to keep the plant in shape.

For vineyards large and small, however, the most widely used method of training is the Guyot system in single or double form. Basically this is a replacement system of pruning whereby the wood which has fruited in the summer is cut out each winter and new wood tied in to fruit in the next summer. The Single Guyot has one fruit carrying arm, whereas the more popular Double Guyot has two. The latter is preferred by most growers as it takes fewer plants for the same amount of space. There is often a place for the Single Guyot where there is not room for two arms. There is no appreciable difference in yield per foot run between the two pruning systems.

In either form, the arms are trained fairly close to the ground to take full advantage of the reflected and radiated warmth from the soil, which is important in our northerly climate.

Support

Vines grown in rows in the open require a stout fence of posts and wires. For the Guyot system, the posts should be spaced 13½ to 15 feet apart (4–4.5 m) depending upon the spacing of the plants within the row.

The end posts should be 7 feet (2 m) long, by 2¾ to 3 inches diameter (7.5 cm), and 7 feet (2 m) by 2 inches diameter (5 cm) for the intermediate posts, driven 1½ feet (45 cm) into the ground.

The wood must be preserved against rot, preferably impregnated under pressure with a preservative or dipped. The fence should be strutted at each end. Galvanised fencing wire is used for the wirework. The two lowest wires are single and the upper wires double. Secure the wires to the end posts with straining bolts and to the intermediates with staples. Do not drive the staples completely home, but allow the wires to run freely through them. The two lower wires (12 gauge/2.5 mm) are set at 15 inches and 21 inches from the ground (40 cm and 55 cm) and the upper wires (14 gauge/2 mm) at 3, 4 and 5 feet (90 cm, 1.2 and 1.5 m) (see p. 47). Preferably the rows should run north to south to reduce mutual shading. If one side of each of the three sets of double wires can be taken down at the time of winter pruning, the task is made easier. To do this, fix a short length of chain to the ends of the wires and instead of wire staples on the intermediates, use small cup hooks. The chain is pulled tight around the end post so that the projecting end of the straining bolt is caught in one of its links. The wire itself is not attached to the bolt (see below).

Spacing

Space the vines 4½ to 5 feet apart in the row (1.3–1.5 m). Allow the wider spacing on good soils. The spacing between the rows is 5 to 6 feet (1.5–1.8 m). If a grass sward is preferred, then space the rows 6 feet apart (1.8 m) with a clean grass-free band of 18 inches wide (45 cm) maintained down the row.

Planting

The young vines should be planted to a stout cane 6 feet (1.8 m) in height, or to a supporting fence post. Plant to the same depth as the vine was in the nursery, being careful not to bury the graft union under the soil's surface where grafted plants are used. The rootstock is resistant to *Phylloxera*, whereas the vine (scion) is not. Burying the union will remove the plant's immunity to the pest and scion rooting might change the vigour of the vine.

A simple method of securing wire at the end using a chain link over the bolt.

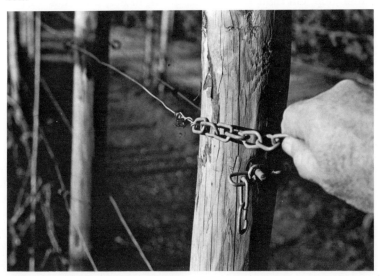

Initial training

Normally, the vine will break into growth some time in May. In the first summer after planting allow the strongest shoot, usually the topmost, to grow and train it vertically up the cane, all others being cut out. In November prune this shoot to a bud just below the bottom wire ensuring that there remain at least two more good buds beneath it. In the second summer train in three shoots, tied vertically, pinching back any others to one leaf. At the end of the second summer after leaf fall, start training to the Double Guyot system. One shoot is tied down to the left and one to the right onto the bottom wires. The third shoot is cut to three buds and this will provide the three replacement shoots for the next year (see photographs). The arms should not overlap with their neighbours, but should be cut to a bud short of the arms on the neighbouring vine.

A weak plant with growth of less than pencil thickness when planted is best left unpruned for a year, and then trained and pruned as above.

Cropping. In the first year of fruiting, two or three bunches only should be allowed on a 3-year-old vine, four or five in the next year, and full cropping thereafter, provided that growth is healthy and vigorous enough to sustain the crop.

Feeding and watering. In February apply Growmore fertilizer at 2 oz per sq yd (66 g/m²) plus sulphate of potash at ½ oz per sq yd (15 g/m²) along each side of the row to a width of 1 foot (30 cm) on either side. On light soils in the spring mulch the vines with a dressing of mushroom compost, peat, well rotted manure or garden compost to a depth of about 2 inches (5 cm). Vines are prone to a magnesium deficiency, sometimes shown by yellowing of the leaves in summer, but not to be confused with chlorosis due to alkaline conditions. If magnesium deficiency symptoms occur, apply a foliar spray of ½ lb magnesium sulphate (coarse Epsom salts) in 2½ gallons of water (220 g/11 litres) plus a few drops of mild liquid detergent and repeat the spray 14 days later. Thereafter apply the magnesium sulphate as a top dressing in the late spring at 2 oz per sq yd (66 g/m²).

Top left: A young vine at the end of the first summer, grown as a single stem.
Top right: The young vine in the autumn is cut to a bud just below the bottom wire.
Bottom left: The young vine; pruning completed.
Bottom right: The vine in the autumn of the second year – three upright shoots have been allowed to grow.

Top left: Pruning the established vine grown on the Double Guyot system. Left hand arm which has fruited is cut back to the replacement shoot, or point of origin.

Top right: Right hand arm which fruited last year is also pruned out.

Below left: The three replacement shoots are left: one arm is tied to the wire at the left, and one to the right.

Finally, the third shoot is cut back to three buds, each of which will produce a replacement shoot next summer.

Part of a multiple vertical cordon (see page 37).

Grapes grown for dessert will benefit from liquid feeds high in potassium in the growing season. Apply the fertilizer once a fortnight from the time the berries start to form until they begin to ripen.

Well established vines are usually able to cope with dry weather except on very shallow soils, nevertheless it is advisable to water them in drought conditions.

Pruning the established vine: opposite

Top. November. Prune as soon as possible after leaf fall, but no later than the end of January. However, early winter pruning stimulates earlier bud burst (by about 7 days) than vines pruned in the spring. Possibly where vines are planted in areas prone to spring frosts, there may be an advantage in pruning in March or April in the hope that they may escape frost damage, but it must be noted that late pruned vines can 'bleed' badly from the pruning cuts. Nevertheless, provided the vines are healthy, the bleeding does not appear to weaken the plants and the flow stops once the plant has developed some leaves. Note, there are, or should be, three strong replacement shoots trained up the cane or post. If not, utilise three strong laterals, nearest to the centre. The two arms which bore fruit last summer are cut back to the replacement shoots. One shoot is then tied down to the left and another to the right arching over the lower wires and the third is cut to 3 or 4 buds. These buds will provide the replacement shoots for next year.

Centre. January. Winter pruning completed. Replacement shoots tied on lower wire and immature wood cut off, leaving 2 to $2\frac{1}{2}$ feet each side (60–75 cm). Third shoot cut to 3 or 4 buds.

Below. July, August, September. The fruit carrying laterals are trained through the double wires. Prune them to 2 or 3 leaves above the top wires as, and when, necessary, using secateurs or shears. Remove any sub-laterals. The replacement shoots (3 for a Double Guyot, 2 for a Single Guyot) are trained up the cane or post and stopped at about 5 feet (1.5 m). Sub-laterals are pinched to one leaf, and as a counsel of perfection, any blossom on the replacement shoots should be removed, though this is not strictly necessary if the growth is strong. Remove any surplus shoots or suckers coming off the main trunk below the two arms. Starting in early September, gradually remove a *small* proportion of the lower leaves, just sufficient to expose the bunches to more sunlight and improve air circulation. Be careful not to expose the bunches too suddenly to the sun as this may lead to sun scorch. This reduction of foliage will help in the ripening of the berries and the control of botrytis.

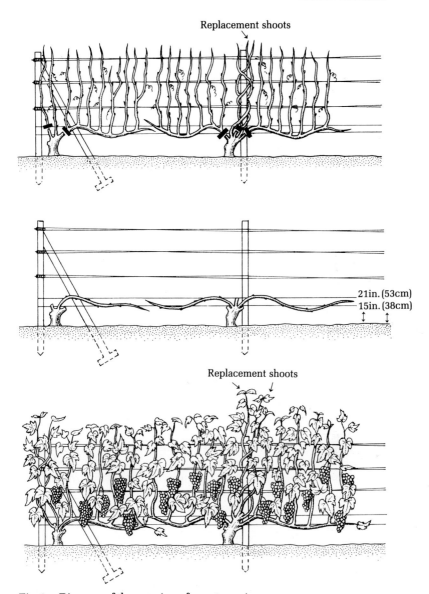

Fig. 6. Diagram of the pruning of a mature vine.

Spray programme

The two most important diseases on grapes grown out of doors are mildew (p. 60) and botrytis (p. 59) and a regular spray programme to control them is advised. Details of pests and diseases which may attack outdoor vines are given on pp. 55–64.

47

The two groups of sprays that should be applied as a routine are as follows:

In mid-June: systemic fungicide, wettable sulphur or sulphur dust for powdery mildew control. Repeat 4 times at fortnightly intervals. (N.B. the minimum interval between spraying and harvesting is 7 days).
Just before the flowers open: systemic fungicide for botrytis control. Repeat every 14 days until 3 weeks before harvest.

Weed control in the open

It is important to control weeds in the first year of planting when the vine is trying to become established as well as in subsequent years when the plant is bearing a crop. Weeds compete for water and nutrients and also are liable to create humid conditions around the bunches making fruits more prone to grey mould. A fairly simple herbicide programme can be adopted using the weed killers simazine and paraquat. Simazine is applied to the soil and is taken up by the germinating weed seeds before emergence. It will not kill weed seedlings that have already emerged nor established weeds. It is used to keep clean land free from weeds. This herbicide creates a seal or barrier to germinating seeds over the surface and the best results are obtained by applying it when the soil is moist, firm and has a fine tilth. It is important to observe the manufacturer's rate of application which should be varied according to whether the soil is light, medium or heavy. Simazine is persistent and the effect should last for about six or seven months. Over-dosing can render the soil sterile and damage the plants. Simazine is applied as an overall spray; the established fruit plants, if wetted, will not be harmed provided the recommended dosage is used.

Paraquat will kill any green tissue, green bark, green buds or leaves on contact. It has a burning or scorching action and should be directed onto the weeds, but must be kept clear of vines. Paraquat will kill annual weeds, and perennial weeds that have no underground food reserves.

Apply the simazine to clean, moist soil in February or March and this should last until the second application after cropping in October. Should any weeds appear in the row during the growing season, these can be killed by a carefully directed application of paraquat avoiding the foliage of the vines.

In no circumstances use the hormone based weed killers on or near vines as these are liable to cause severe distortion.

Cultivars for growing outside

Black grapes, other than a few hybrids, do not ripen sufficiently well when grown outside to make a good wine, except in hot summers. They are best grown on a wall or fence or given some protection for at least part of the season. For this reason in most commercial vineyards in Britain, white grapes predominate. In general, white cultivars which ripen in early to mid October in the south are preferred as the early ones are very liable to be attacked by wasps. Further north early cultivars should be chosen rather than the late ones which might not ripen.

All grapes can be made into wine, but there are a few outdoor cultivars suitable for dessert, given a good summer, and these are mentioned below. Nevertheless, as might be expected, they will acquire a sweeter flavour if grown against a warm wall. Cultivars normally recommended for under glass are unsuitable.

The season of ripening given below refers to the south of England.

WHITE

Chasselas (late). Of which there are a number of clones, e.g. Chasselas d'Or, Chasselas 1921, Chasselas Rose Royale and Royal Muscadine. All yield good quality berries excellent for dessert and wine: however, this cultivar is too late in the open and is better on walls. Late October.

Madeleine Angevine 7972 (early). A heavy cropper. Vigorous. Prone to mildew. Relatively hardy. Suitable for colder areas. White wine. Quality only fair. Late September–early October.

Müller Thurgau (Riesling Sylvaner) (mid season). The most widely planted. It has a delicate flavour and gives an excellent wine. Passable as dessert. Needs good weather during pollination. Mid October. Highly recommended.

New York Muscat. A tawny coloured grape of excellent muscat flavour in a hot summer. Must be grown on a wall. Large berries. Mid to end October. Dessert.

Perle de Czaba (early). An excellent muscat flavoured berry. Best on a wall or under glass. Weak to moderate vigour. Light cropper. Late September. Dessert.

Precoce de Malingre (early). A large berried sweetwater grape of good flavour. Moderate vigour. Light cropper. Dessert and wine. Late September.

Seyval (Seyve Villard 5.276) (mid-season). A hybrid, with some resistance to powdery and downy mildew. A heavy cropper and easy to grow. Blends well with Muller Thurgau. Recommended. Mid to end October.

Siegerrebe (very early). A golden brown berry of good flavour with a trace of muscat and high sugar content. Prone to wasps. Dessert and wine. Medium vigour. Late August—early September.

BLACK

Brant (late). A hybrid of Canadian origin often grown for its autumn colour, bears heavy crops of small, sweet black grapes, liked by children. Some resistance to mildew. Vigorous, useful wall cover. Mid October.

Cascade (Seibel 13.053) (mid-season). Resistant to mildew. Low in sugar and high in acid. A fair quality wine. Small bunches of deep purple berries with red juice. Very vigorous, excellent as a wall cover plant. Early October.

Opposite: 'Seyval', a good hybrid which is easily grown outside in Britain. Below: 'Müller Thurgau' grown on a wall.

Leon Millot (mid-season). A French hybrid with good resistance to mildew. Small bunches of blue-black berries. Very vigorous. Makes a fair quality red wine. Crops well. Reliable. Early October.

Muscat Bleu. A muscat flavoured berry of good size. Bunches medium. Excellent flavour but seeds are large. Best on a wall, but will ripen out in the open in a warm summer. Attractive autumn colour. Mid October. Dessert.

Noir Hatif de Marseilles (early). Small to medium sized berries and bunches of slight muscat flavour. Good quality. Rather a weak grower and requires fertile soil. Needs protection or a warm wall. Late September. Dessert.

Pirovano 14. Medium sized bunches of large berries of good non-muscat flavour suitable for dessert and wine. Best on a wall or with protection. Mid October.

Tereshkova (early). A hybrid with medium sized berries of purple-red with delicate muscat flavour. Attractive autumn colour. Late September. Dessert.

'Cascade', a vigorous black vine, ripening in mid-season.

AMERICAN VINES

These are North American hybrids often with the vine species *Vitis labrusca* in the parentage. There are a number of cultivars e.g. Concord (black) and Fragola (pink). Small or large bunches of good sized berries, having a thick slippery skin and a taste faintly reminiscent of strawberries. Sometimes described as having a foxy flavour which is imparted into wine made from them. Best grown against a wall. Good resistance to mildew. Late October.

PROMISING CULTIVARS

Bacchus. White. Similar to Müller Thurgau, but more reliable in cropping. Gives a good quality wine.

Huxelrebe. A white grape, producing a good quality muscat flavoured wine slightly earlier than Müller Thurgau. Susceptible to mildew. A heavy cropper. Early October.

Reichensteiner. A white grape, high in sugar, but makes a neutral wine. Crops heavily. Some resistance to botrytis. Early October.

Scheurebe. A white wine grape. Mid-season, similar to Müller Thurgau, but a heavier cropper. Mid October.

Schonburger. A white grape which makes a good quality wine. Ripens the same time as Müller Thurgau, but requires a warm site to crop well.

ROOTSTOCKS

Most vines supplied by nurserymen are on their own roots and these are satisfactory provided *Phylloxera* does not become a problem in this country. Imported vines are grafted onto rootstocks which are resistant to this pest and some have the added advantages of influencing vigour and being more suited to certain soil conditions as follows:

Suitability for soil types					
Shallow poor, stony, dry	Deep fertile without chalk	Deep fertile with chalk	Heavy clay with chalk	Heavy clay	Chalk
5BB 125AA	5C SO4 125AA	5C SO4 5BB 125AA	SO4	125AA 5BB SO4	SO4

Above: 'Siegerrebe', a very early grape, suitable for dessert and wine.
Below: 'Tereshkova', a hybrid which produces attractive autumn colour.

Pests*

Grapes grown out of doors are generally much less liable to be attacked by pests troublesome in greenhouses such as mealybugs, scale insects, red spider mite and whitefly. However, an outdoor vine growing in a warm sheltered position, such as against a south facing wall, may also have these problems. When applying pesticides there is a danger that the bloom on glasshouse grapes will be marred if sprays are applied after the fruits have begun to swell. As far as possible, pests should be controlled before this stage is reached, or formulations less likely to affect the bloom, such as smokes or aerosols, should be used. Insecticides should not be used within 10 days of picking the fruit.

The first five pests listed below all have a wide range of host plants. If the vine is being grown in a greenhouse with other plants, these should also be checked for the presence of pests and treated accordingly.

Birds. Birds are the most damaging pests of outdoor grapes and they may eat the entire crop unless the vine is covered with bird-proof netting. This will need to be in place from the beginning of September to the middle of October depending on the cultivars grown. Netting with a mesh of ¾-inch or an inch (19–25 mm) must be used. If the vine is too large to be completely netted, individual bunches can be enclosed in bags made from muslin or nylon stockings.

Glasshouse red spider mite. Red spider mites are tiny, eight-legged, sap-feeding animals that occur in large numbers on the underside of the leaves. They overwinter as adult mites and they begin feeding on the new foliage during April or May.

The first sign of an attack is a fine yellowish green speckling that can be seen on the upper leaf surface; examination of the underside of the leaf with a hand lens should reveal the mites. As the infestation increases, leaves begin to dry up and drop off, and the vine may become covered in a fine silken webbing which is produced by the mites. Despite their common name, the mites are not really red and they range in colour from nearly black to yellowish green or orange.

* by Andrew Halstead, entomologist at RHS Garden, Wisley.

Early treatment is required to prevent a damaging infestation from developing. As soon as mite damage is seen, spray the vine with malathion, repeating at intervals as necessary. As an alternative to using pesticides, red spider mite can be controlled by introducing a predatory mite called *Phytoseiulus persimilis*, although this cannot be used if insecticides are being used to control other pests. Supplies can be obtained from: Natural Pest Control (Amateur), Watermead, Yapton Road, Barnham, Bognor Regis, West Sussex PO22 0BQ; Bunting and Sons, The Nurseries, Great Horksley, Colchester, Essex.

Glasshouse whitefly. Adult whiteflies are tiny moth-like insects with white wings which live on the underside of the leaves. The nymphal stages are flat, whitish green scales, and both the adults and nymphs feed by sucking sap. Like mealybugs, whitefly can create problems with honeydew and sooty mould. The eggs and nymphal stages are relatively immune to insecticides, so frequent sprays are necessary to kill the adults as they develop.

Sprays or smokes containing permethrin or a related compound, such as pyrethrum, are effective. As an alternative to insecticides, a parasitic wasp called *Encarsia formosa* can be introduced during the summer. This is available from the same sources as *Phytoseiulus* (see above).

Mealybug. This is a common pest of glasshouse grapes. Mealybugs are soft bodied, pinkish grey insects up to $\frac{1}{8}$-inch (3 mm) long, and they suck sap from the leaves and stems. They tend to cluster in the leaf axils and they secrete white woolly fibres around themselves. They excrete a sugary substance called honeydew that makes the vine sticky and encourages the growth of a black sooty mould on the surface of the foliage and fruits.

The winter tar oil treatment described below under scale insects also deals with mealybugs. If they persist during the growing season they can be controlled by thorough spraying with malathion at 14-day intervals during the spring and early summer.

Scale insects. Scale insects feed by sucking sap from the vine and the type most commonly found is called brown scale. The mature insects are covered by hard, shiny brown, convex shells about $\frac{1}{4}$-inch (6 mm) long and they are attached to vine stems. Apart from the recently hatched nymphs, scale insects do not move once they have found a suitable feeding place.

A less common but more spectacular type of scale insect is the woolly vine scale. The mature females of this scale have dark brown, wrinkled shells about $\frac{3}{16}$-inch (4 mm) in diameter, which are perched on the edge of a mass of white waxy threads which contains the eggs.

Scale insects can be controlled by thoroughly brushing the vine rods with a tar oil wash during December. Tar oil will damage any plants that are in leaf, so care must be taken if other plants are growing nearby. Before treatment, the vine rods should be scraped to remove loose bark and as many scales as possible. Collect and burn the scrapings. If necessary, the vine can also be sprayed with malathion in late June and mid July as the young nymphs begin to hatch.

Vine blister mite (Erinose). This is a microscopic mite that can occur on both glasshouse and outdoor grapes. It overwinters inside the buds and starts feeding on the leaves in the spring. As a result of this feeding the upper surface of the leaves develops raised patches and the underside of these blisters is covered by fine white hairs. These hairs may be confused with mildew disease but the blistering effect is diagnostic for the mites. The symptoms first appear in May or June, and as the summer progresses the hairs darken and become reddish brown. The mites live among the hairs and feed on the leaf surface but, apart from some disfigurement to the leaves, they seem to cause no real harm. Infestations can usually be dealt with by picking off the affected leaves.

Vine phylloxera. This serious pest of vines is rare in this country although outbreaks occasionally occur when infested vines are imported into Britain. It is an aphid-like insect with a complex life cycle, having different forms that cause galls on the leaves and roots. The leaf galls are the most obvious symptom. They are pinkish or yellowish green, swollen, rounded structures that form on the leaf surface. The galls are hollow and contain the pest. The root infesting forms cause swellings on the roots and heavy infestations will kill the plant unless the vine has been grafted onto a phylloxera-resistant rootstock. Suspected outbreaks must be reported immediately to the local office of the Ministry of Agriculture who will take the appropriate control measures.

Vine weevil. The adult weevils feed at night and eat irregular shaped notches from the leaf margins of many plants, including grape vine. The beetles are mainly black with small brown patches on the wing cases. The slow moving beetles are about ⅜-inch (9 mm) long and they have elbowed antennae. Their larvae are fat, creamy white, legless grubs with brown heads. They live in the soil and feed on roots. Well established vines are unlikely to

57

suffer serious damage, but pot-grown or young vines may be killed by the grubs. If damage caused by the grubs or adults is seen, they can be controlled by drenching the soil with a spray strength mixture of HCH (BHC).

Wasps. In some years wasps are very numerous and they will damage large numbers of grapes, especially of the early ripening varieties. An effort should be made to find and destroy as many wasp nests as possible in the locality. It is easier to follow the wasps' flight paths as the sun is setting and likely places to look for nests are ditches, hedge bottoms, and under the eaves of buildings. Once located, the nests can be destroyed by placing a wasp powder such as carbaryl in the entrance hole. Individual bunches of grapes can be protected from wasps by enclosing them in bags made from muslin or old nylon stockings. Wasps can also be kept out of greenhouses by screening the ventilators with nylon netting.

Left: Fruit protected from wasps by an old nylon stocking.
Right: Net supported over vines to protect fruit from bird damage.

Diseases and disorders*

DISEASES

Downy mildew occurs only occasionally in this country, mainly on outdoor grapes. It shows as light green blotches on the upper surface of the leaves and as a downy mildew on the lower surface of these patches. Affected areas dry up and become brittle causing the leaves to curl and fall. Diseased berries shrivel and become brown and leathery. The tips of the shoots may also be attacked. Remove and destroy all diseased tendrils and leaves to remove the overwintering stage of the fungus, although some spores also overwinter in the bud scales and shoots. Where infection is expected to occur, apply a protective spray of zineb, mancozeb, bordeaux mixture or liquid copper, *not* when the vines are in flower, and repeat at 10 to 14 day intervals, ceasing when further spraying would cause an unsightly spotting of the fruit. (See also Vine Blister Mite, p. 57).

Grey mould (botrytis) is the most troublesome disease on outdoor vines, but it can also be a nuisance on indoor grapes in poorly ventilated greenhouses where humidity is high.

Affected grapes rot and become covered with a dense brownish grey furry mass of fungal growth. The fungus either attacks the berries through wounds or it may invade the floral parts so that some of the fruits are already diseased as they develop. By whichever method infection occurs, once the disease is established it can spread rapidly by contact and also by means of air-borne spores which are produced in vast numbers. In wet weather and in very humid greenhouses crop loss can be considerable. Under glass the trouble can be prevented to a certain extent by adequate ventilation to reduce the humidity and by the prompt removal of unhealthy berries and leaves. Once the disease has occurred fumigate the greenhouse with tecnazene smokes.

It is difficult to control the disease on outdoor grapes so in wet seasons try to improve the aeration around the bunches by thinning out some of the berries and judicious removal of some of the shoots. Sprays of benomyl, carbendazim or thiophanate-methyl will control grey mould for a season or two, but regular

* by A. V. Brooks, plant pathologist at the RHS Garden, Wisley

use of these fungicides may lead to the development of strains of the fungus which are resistant to them so that they will become ineffective. For those growing grapes on a commercial scale, fungicides are available to control this disease, and advice should be sought on this subject.

Honey fungus. Both indoor and outdoor vines are very susceptible to infection by this soil-borne fungus which can kill affected plants very rapidly. White fan-shaped growths of fungus develop beneath the bark of the roots and main stems at and just above ground level. Brownish black root-like structures known as rhizomorphs may be found growing on diseased roots; these grow out through the soil and spread the disease. Control of the fungus is difficult and it is essential, therefore, to trace the source of infection so that all woody debris can be dug up and burnt together with dead and dying vines and as many roots as possible. The soil should be treated with a proprietary product containing a phenolic emulsion or sterilised with a 2% solution of formaldehyde or should be changed before replanting.

Powdery mildew is most troublesome on indoor vines particularly in a cold greenhouse especially if the soil is dry and the atmosphere is humid or stagnant. Powdery mildew is also common on outdoor grapes, particularly on those growing in very dry positions such as against walls. A white powdery coating of fungus spores develops only sparsely and the most obvious symptom is a grey or purplish discoloration of the diseased areas. The disease can also attack the flowers and fruits causing them to drop or at a later stage the grapes may become hard and distorted and split, and are then often affected by secondary fungi, such as grey mould (see p. 59), which cause extensive rotting.

Powdery mildew can be prevented to a certain extent by mulching and watering to stop the soil from drying out. As soon as the disease appears apply benomyl, thiophanate-methyl or triforine with bupirimate; spraying or dusting with sulphur can also be carried out. If the disease has been troublesome in previous years the first application of fungicide, whichever is used, should be given 10 to 14 days before mildew is expected. Four applications during the season may be needed to keep the disease in check. In a cold greenhouse in dull weather, it may be necessary to provide some heat temporarily so as to avoid too humid an atmosphere and careful but sufficient ventilation should be given to get good air circulation. Over-crowding of the shoots and leaves must be avoided to prevent stagnant air conditions. In a warm greenhouse the pipes can be painted with a sulphur solution but this must be done with care to avoid contact with and scorching of the lower foliage. The solution is made by mixing well equal parts of flowers of sulphur and soft soap, making lumps about the size of golf balls. One of these is put into a jam jar with a little water and stirred with the brush which is used to paint the pipes. The same mixture can be used to paint the vine stems while they are dormant in the winter, after the loose bark has been gently scraped away with a knife (see p. 12).

DISORDERS

'Bleeding' of the wounds. Often as a result of late winter pruning, the sap flows freely from the pruning cuts, especially if the wounds are large. The sight of such 'bleeding' can be disconcerting to the inexperienced grower. There is no need for concern, however, as provided the vine is healthy, the sap loss does not appear to harm the plant and it will eventually stop in the spring when the buds break into growth. There have been many remedies advocated in the past to staunch the flow such as cauterising the wound, applying sealing wax, carpenter's knotting etc., but these rarely work once the 'bleeding' has started. The real answer is not to prune late, but to complete the job by the end of December, and, preferably, immediately after leaf fall.

Exudations. A common phenomenon in the spring is the presence of transparent globules resembling eggs, on the lower leaf surfaces and petioles. These small round greenish or colourless droplets are not the eggs of any pest but are due to a natural exudation from the plant. This type of exudation commonly occurs on young growth and indicates that the root action is very vigorous and the plant is in good health. The symptoms are inclined to be more obvious, however, on plants growing under glass where the atmosphere is humid.

Fruit splitting. This trouble most commonly occurs as a result of an attack by powdery mildew (see p. 60). However it is occasionally due to irregular growth such as may occur when heavy rain follows a period of drought. Affected grapes usually rot as a result of subsequent grey mould infection. There is very little which can be done to prevent this trouble apart from mulching to conserve moisture and by watering in dry periods before the soil dries out completely.

Magnesium deficiency. Vines are soon affected by a soil deficiency of magnesium, symptoms being produced in the leaves. These usually show as a yellowish orange discoloration between the veins, but in some cultivars the leaves may have purplish blotches. Later the affected areas turn brown; these symptoms should not be confused with sun scorch (p. 64). The deficiency can be corrected by spraying the foliage at the first signs of trouble, with ½lb (220 g) of magnesium sulphate in 2½ gal (11 litres) of water plus a spreader such as soft soap or a few drops of a mild washing-up liquid. Apply two or three times at fortnightly intervals. Soil applications act more slowly, but are useful as a long-term cure. For outdoor grapes apply the magnesium sulphate in subsequent years as a soil dressing in late spring at 2 oz per sq yd (66 g/m²).

Oedema is a physiological disorder which is more likely to occur on vines grown under glass. Fairly late in the season, small wart-like outgrowths appear on the stalks when the fruits are developing and sometimes on the grapes themselves and even on the lower leaf surfaces. These outgrowths may break open and then have a blister-like or white powdery appearance, or they may become rusty coloured and appear as brown scaly patches. The trouble is caused when the roots of an affected plant take up more water than the leaves can transpire and this may be due to extremely moist conditions either in the soil or in the atmosphere or both. Once oedema has occurred, do not remove the affected

Below: Magnesium deficiency symptoms.
Opposite: Powdery mildew (see page 60).

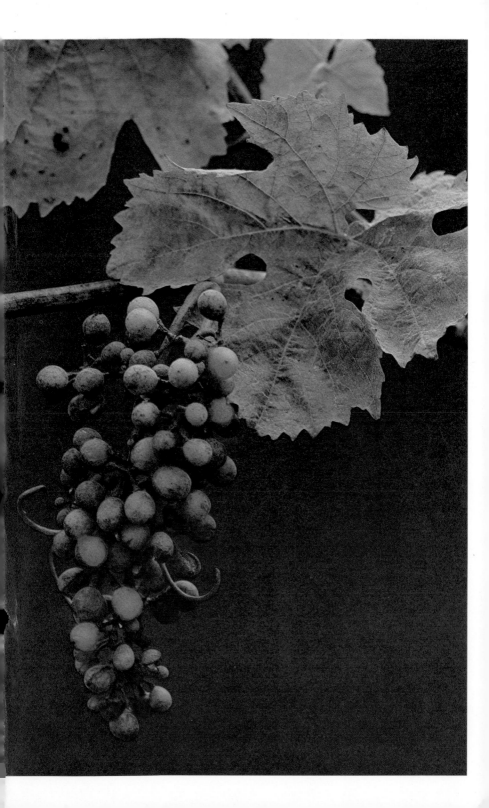

parts as this will only make the trouble worse. No special treatment can be given and the remedy is to maintain drier conditions both in the air and soil; with correct cultural treatment the affected plant should recover in due course.

Scald and scorch. Scald which shows as discoloured sunken patches on the berries, and scorching of the foliage which results in large pale brown patches, is due to hot sun striking through glass on to moist tissues. These troubles usually occur where the ventilation has been poor, but light shading may also help to prevent them. Remove affected berries and leaves.

Shanking first shows at the early ripening stage when odd berries or small groups of berries do not colour and develop naturally. It starts as a dark spot along the stalk of the grape which is finally girdled so that it shrivels and the grape stops developing and fails to ripen. The grapes are watery; black grapes turn red and white grapes remain translucent. The fruit if tasted will be found sour and unpalatable.

This disorder is usually a result of one or more unsuitable cultural conditions, such as faulty root action, due to too wet or too dry soil conditions, or the penetration of stagnant soil by the roots, or over-cropping of the vine which puts an undue strain upon the root system. The remedy, therefore is to study the soil conditions to see whether the roots could have been affected by drought or waterlogging. Mulching or resoiling may be necessary if there is any sign of such damage. At the same time reduce the crop for a year or two until the vine has regained its vigour. When shanking occurs fairly early in the season it is sometimes possible to save the rest of the crop by cutting out the withered berries and spraying the foliage with a foliar feed, providing the vine is not being overcropped by taking too many bunches.

Spray damage. Damage by sprays, usually due to the misuse of weedkillers (such as 2,4-D; 2,4,5-T; mecoprop), occurs on both indoor and outdoor vines. Affected leaves become narrow and fan-shaped, are frequently cupped and the shoots twist spirally. Affected plants usually grow out of the symptoms in due course, but it is better to avoid such damage by the careful use of hormone weedkillers. Keep special equipment for their use, do not spray on a windy day nor use within 400 yards of outdoor grapes. Close all ventilators in greenhouses if any spraying is to be carried out in the vicinity. Do not leave weedkillers in a greenhouse as vapours from them may affect plants on a hot day.